THE BUSINESS OF LIVE STREAMING

4 C's to Expanding Your Reach

BY

SHERRY BRONSON

Watersprings PUBLISHING

Published by Watersprings Publishing, a division of
Watersprings Media House, LLC.
P.O. Box 1284
Olive Branch, MS 38654
www.waterspringsmedia.com
Contact publisher for bulk orders and permission requests.

Printed in the United States of America.

Library of Congress Control Number: 2020916728

ISBN-13: 978-1-948877-60-2

TABLE OF CONTENTS

INTRODUCTION

Welcome to the Digital Age! This live streaming guide is for small businesses, communicators, authors, trainers, consultants, speakers, educators, and other professionals. This book was being written during the COVID-19 crisis. We all understand now more than ever that live video streaming is critical in a pandemic such as COVID-19. We watched the world stand still as the world economy was closed for business. Most industries had to transition primarily into online services, To-Go and delivery only, and quickly re-invent their business model. Teachers had to learn how to instruct their students online. Parents were very focused on helping with their children's online education. Some students in rural and low-income families in America still had little to no access to internet. This tragedy quickly transitioned everyone to rely on live streaming and thus validating its importance.

Live streaming is a powerful tool. It is essential that business owners include it in their online video marketing toolkit. My goal is to eliminate the feeling of being overwhelmed and worrying about breaking the bank.

This book serves as a valuable guide for the budget conscious small business owner who is seeking solutions for lowering marketing costs and promoting their services through video. With the integration of the internet, small businesses have a tremendous opportunity to increase revenue using live streaming, website, and social media, Facebook and audience engagement. Live streaming also allows businesses to build new relationships with customers in real time.

This interconnected global world is mandating that we reset our thinking, challenging us to learn new technology, execute new concepts and ideas using both the internet and live streaming. Customers are requiring businesses of many industries to provide their products and services much faster.

Is your mindset still in the analog era or have you joined the digital media age? I implore you to use your creativity and learn live streaming as another platform to engage with your customers in a more consistent manner.

Small business owners should be encouraged about live streaming. Live streaming using Facebook is an affordable financial investment. It will increase your brand awareness and generate new leads.

We should be leveraging Facebook live stream to engage and interact with our customers, educate, and promote our services and products.

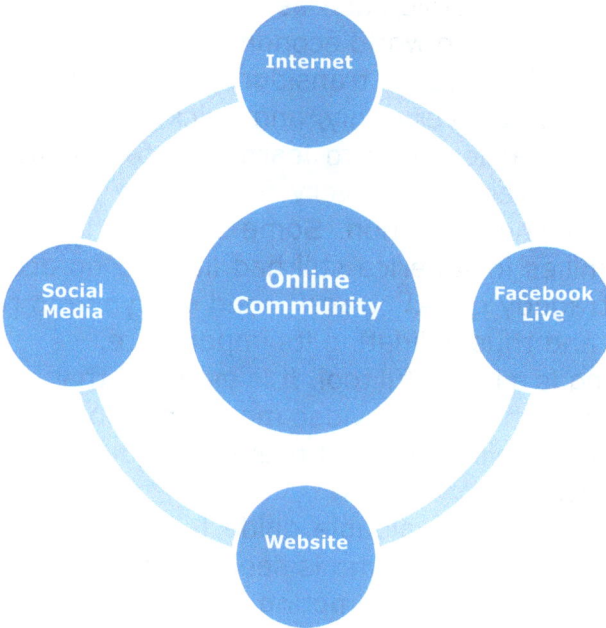

4 C'S TO EXPANDING YOUR REACH

CONTENT, CAPTURE, CONVERT & CASH

BIZ TIDBIT

Small business owners do not despise the day
of small beginnings. The information age has opened
boundless opportunities to reach
your ideal clients via the internet.

SHERRY BRONSON

CHAPTER 1

INTRODUCTION TO LIVE VIDEO STREAMING USING FACEBOOK LIVE

Currently, how are you using Facebook live? Is your business harnessing the power of the internet by using Facebook Live? Live streaming for business is an essential part of an online marketing plan. Live stream video campaigns will definitely generate buzz, and businesses should not be afraid of technology. Facebook live allows businesses and communicators to engage directly with their audience, reduce marketing costs, access to 2 Billion active monthly users on Facebook, and increase brand awareness. Facebook live videos are watched 3 times longer than regular videos. There has become an immense demand for live content and we must be ready to meet this demand.

What is holding you back from Live Streaming? Live streaming is quite valuable and a great opportunity to create a new revenue stream and expand business reach. Business growth can be hindered if live streaming is identified as a nuisance rather than being a useful tool to solve customers' problems. Businesses should embrace the information age at all costs in order to be competitive in the marketplace. Due to the demand of live streaming content, it is inevitable that changes to business practices must take place. There will be certain behaviors that must be unlearned and an assertive pursuit to learn new strategies to incorporate in the business model.

BIZ TIDBIT

81% of businesses prefer to use Facebook
for their video marketing.

SOURCE: BUFFER

**DEFINITION:
LIVE STREAMING**

Transmit or receive live video and audio
coverage of (an event) over the internet.
A live transmission of an event
over the internet.

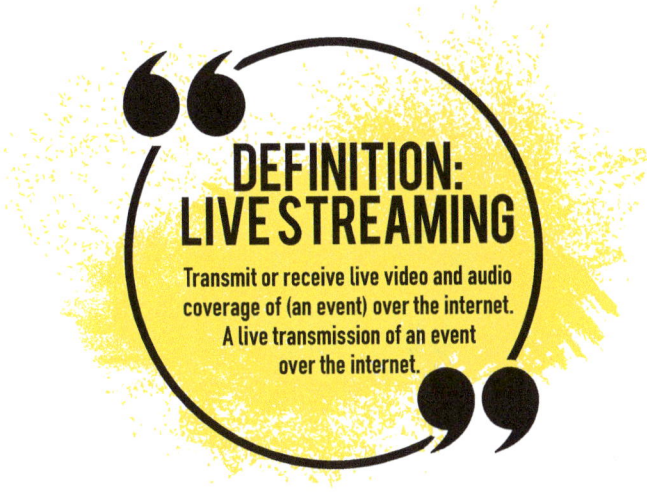

WHAT IS LIVE STREAMING?

DID YOU KNOW?

- FACEBOOK was founded in February 4, 2004.

- FACEBOOK LIVE was launched August 5, 2015.

- During March 2019 Facebook on average had 1.56 billion daily active users.
(https://newsroom.fb.com)

- 2.38 billion monthly active users on Facebook as of March 31, 2019.
(https://newsroom.fb.com)

- Facebook has over 8 billion video views per day. (Source: TechCrunch)

- In 2018, 94% of marketers were using Facebook for advertising.
(Source: Social Media Examiner)

- Until April 2018, there were over 3,500,000,000 Facebook Live broadcasts. (Source: Engadget)

CHAPTER 2

CONTENT

What is content?

Content is the **first "C"** in the business of live streaming. It is the pre-go live or planning phase of your business going live on Facebook or any social media platform. What information do you want to share with your existing and new customers? Are you wanting to promote a product or service? Are you asking for feedback from your clients?

As part of your live video streaming content strategy, please determine the content that you will be sharing on a consistent basis. In this chapter, I will share some definitions and terminology used in digital marketing.

TERMINOLOGY: CONTENT CREATOR

A content creator is someone who is responsible for the contribution of information to any media and most especially to digital media. They usually target a specific end-user/audience in specific contexts. A content creator can contribute any of the following: blog, news, image, video, audio, email, social updates, and other related content. (Source: https://www.stateofdigitalpublishing.com/content-strategy/what-is-a-content-creator/)

According to Facebook, live videos have six times the engagement of pre-recorded videos. Savvy brands will implement live streaming across mobile apps and social platforms such as Facebook, Instagram, and Snapchat. (Source: https://www.stateofdigitalpublishing.com/insights/mobile-video-consumption/)

LIVE STREAMING CONTENT IDEAS

Here are some examples of content ideas:

- Live Product or Service Demonstration such as applying make-up, install or repairs, applications (apps), fitness, etc.

- Live Product Promotions such as new books, jewelry, new technology, etc.

- Live Teaching: Webinar or Training Sessions

- Live Interviews with Business Expert Panels, Author Showcases, Celebrities, and Athletes

DEFINITION: CONTENT

Content is both information and communication.
BUSINESSDICTIONARY.COM

- Get to Know Me (Question & Answer Live)

- Live Talk Show or Web Series

- Live QVC Style Fashion Show

- Business Grand Opening: Livestream Discount Sales promo offer for first time customers

- Live Events, Conferences, Church Services

- Live Unboxing of a new product to share with your clients

DIGITAL AND TRADITIONAL MARKETING

Out with the old and in with the new – not so fast. Yes, times have changed, and we must change with it. We can sometimes be too quick and dismiss traditional marketing strategies that were very successful.

Digital marketing alone is not always a winning strategy. Our ultimate goal is to reach our niche or target audience and help solve their problems and meet their needs. We are looking for our tribe and our tribe is looking for us. In order to reach our ideal customers, this will require business owners to launch out and engage in the boundless virtual world called the Internet. The combination of digital media such as the internet and Facebook live will boost your opportunity to reach your ideal customers.

In this chapter, I listed both traditional advertising and marketing tools that will help you determine what works best for you.

Traditional Marketing	Digital Marketing
Direct Mail (postcards, catalogues, etc.)	Email Marketing
Face-to-Face Networking	Virtual Online Marketing
Signage	Digital Signage
Telemarketing: Cold Calling	Customer Relationship Management (CRM)
Newspaper (print)	Digital Online Format
Television Advertising	Streaming Television Ads

I encourage you to include Facebook Live in your new digital marketing plan. Please conduct your own marketing research to determine if live streaming fits your business marketing strategy.

For those of you who still love traditional marketing, digital marketing has not completely overtaken traditional marketing. In combination, digital and traditional marketing provide small business owners with the unlimited potential to generate new leads. As we have entered a generation shift from baby boomers to millennial, we cannot neglect the fact that human interaction (face-to-face) is still vital in our global society. Face-to-Face and word of mouth marketing are still powerful activities for marketing your business products and services. Human interaction still allows us to connect, collaborate, and build lasting relationships.

Both digital and traditional marketing have advantages and disadvantages. I highly recommend that you do your research and consider all options before investing your time and money in both traditional and digital marketing campaigns. You can always launch your marketing campaigns in phases.

> ## BIZ TIDBIT: DIGITAL MARKETING STRATEGY
>
> Your digital marketing strategy is the series of actions that help you achieve your company goals through carefully selected online marketing channels. These channels include paid, earned, and owned media, and can all support a common campaign around a particular line of business.
>
> HUBSPOT.COM

> ## BIZ TIDBIT: STRATEGY
> Strategy is just a plan of action to achieve a desired goal, or multiple goals.
>
> HUBSPOT.COM

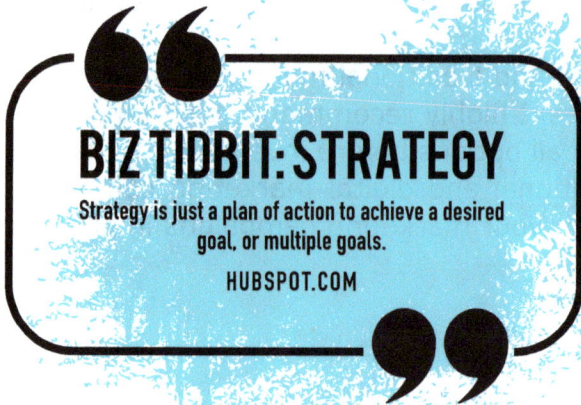

ONLINE VIDEO CONTENT MARKETING

Why Use Video Marketing?

Many consumers have chosen video as a preferred source for information consumption. They are not satisfied with only reading newspapers, magazines, etc. Video is the most powerful digital marketing tool to reach your audience and expand your reach.

Video captures consumer's attention immediately, especially when the quality and content are great.

BUILD CREDIBILITY WITH VIDEO MARKETING

Here are a few video marketing tips for small businesses:

Customer Testimonial Videos
Hire a professional videographer in your area to record and edit the 2-3 positive testimonials about your business. Video should be 2 to 2:30 minutes.

Webinars / Video Training
Use Facebook Live, Zoom.com, or ClickMeeting.com to host your webinars and build credibility.

Video Promos
Excite Your Customers: Use any of these: www.Promo.com, Magisto.com, to create professional marketing videos for your business.

Newsletters
When communicating with your customer, embed your marketing videos in your customer newsletters. Resource: ConstantContact.com, MailChimp.com

Landing Pages
Great for marketing your business: promote your business, sell and promote product and services. Resource: Leadpages.net

BIZ TIDBIT

Live Streaming video content is a powerful new tool in the digital marketing toolbox that no small business can afford to ignore.

SMALLBIZTRENDS.COM

LIGHTS, CAMERA, & SPEAK:

HOW TO BOOST YOUR CONFIDENCE SPEAKING ON FACEBOOK LIVE.

As a person who enjoys speaking on-camera to my audience, I must say it takes practice and more practice. I have to be transparent with you about overcoming the fear and nerves in the belly to speak confidently on-camera. When I first started public speaking, my heart would be racing, my hands would shake, and my legs would tremble slightly. Of course, I had rehearsed my message and was very familiar with the content that I was about to deliver. Once I would get through the introduction of the message, I would start to relax and delivered my message with clarity and positive impact. By speaking consistently, I eventually overcame the fear and all of the nervous symptoms diminished.

There are some people who speak naturally on camera and there are others who will require practice, which is absolutely fine. This new digital era is challenging us to conquer our fears, forgive our flaws, and find our voices to engage directly with our audience. But there is good news for the camera-shy entrepreneur, you do not have to be perfect on-camera. In the age of reality television, people are fine with you being your authentic self. One of the keys to customers staying engaged is to be knowledgeable about the business products and services and how it solves a problem(s) that the customer is experiencing.

BIZ TIDBIT

6 out of 10 people would rather watch online videos than television.

GOOGLE

3 BIZ TIDBITS TO BOOST YOUR CONFIDENCE

Ready, Set, & Go Facebook Live!

1. OVERCOME BEING CAMERA SHY

Does the thought of speaking in front of a live audience give you butterflies in your stomach?

This nervous feeling usually arises when you are afraid of public speaking. This is why practicing your message will be essential before you go live. I am not a fan of winging it, or fake it until you make it.

If your messaging is not genuine and you are not prepared, your target audience will see right through it. Always remember, during your Facebook Live webcast, the audience will scroll pass you or write an unfavorable comment on your live webcast if you are falling flat on delivering your content.

Yes, there are many people who go live on Facebook with their mobile devices without preparation. As a business owner, there is not a real advantage for you to go live on Facebook without having specific content to communicate to your tribe.

Effective communication is critical when you are promoting your products and services on Facebook Live. We usually get one shot at making a lasting impression. As it is said, always make a good first impression because it can make or break you. Once you have decided to host a Facebook Live Webcast, I highly recommend that you write and plan out each week's topics. Preparation and rehearsal will boost your personal confidence and allow you to be more relaxed during your presentation. As you consistently host Facebook Live Webcasts, you will begin to become comfortable enough to host impromptu live webcasts.

One of the most powerful techniques to calm your nerves that actually works is to begin doing breathing exercises before and during your live webcast.

Source: http://sixminutes.dlugan.com/
vocal-variety-speech-breathing/

2. ABOUT YOU: SHARE REAL LIFE STORIES

Yes, people want to know who you are. Most people love good stories of overcoming adversity and success. Even on our business websites, we have a page that says, "About Us." Your first Facebook Live webcast will be your opportunity to introduce yourself to potential leads and customers.

As an entrepreneur, you must be able to be real with your customers. That does not mean to overshare about your personal life. Please keep it professional. I am talking about sharing stories of your lows and highs, growth, triumphs, and customer satisfaction experience with your business. Sharing your stories will allow them to see that you are a real human being who is passionate about building your dreams. Your story will inspire and motivate them to stay connected each week on your live webcast, buy your products and services, and tell others about you and your business. This will give you a competitive edge because you are offering that personal touch to your customers. Happy storytelling to you and build those relationships!

3. KNOW YOUR CONTENT

What information do you want to convey to your tribe? What is the call-to-action? How will this live webcast benefit your clients and potential leads?

We all understand that time is so precious, and we value it. Your Facebook Live webcast must have purpose. The live broadcast plan will have listed what you want to accomplish in 10, 20, 30 minutes or a 1-hour live webcast.

- **BOOKS:** If you are launching a new book, please have an outline of what you want to accomplish to giveaway and/or sell your books.
- **PRODUCTS:** If you are promoting a new product, please have the product available. Share how the product will help solve a problem for them and how they can purchase it during the live webcast.
- **SERVICE:** If you are hosting an upcoming webinar, please during the live webcast, share with the audience what they will learn from attending the webinar. Please share with them how the webinar will help them with their business, personal life, or career. Please give away a freebie for registering.

The Facebook Live webcast must have goals and objectives that benefit the viewer in a good way. If the audience can tell that you are not just selling to them, they are more likely to tune in and hear what you have to say. The content that you want to share must be clear.

With that said, you must know your products and services better than anyone. You are the visionary, go forward and share the vision with the world!

CHAPTER 3

CAPTURE THE CONTENT

Are you using Facebook Live as a hobby? Are you ready to expand your reach and grow your business? Live streaming for businesses should be treated as a real opportunity to share your products and services, engage with potential customers, and generate leads. Small business owners cannot afford to use Facebook Live as a hobby. In this chapter, I will provide the information needed to help businesses determine the level that you would like to get started with live streaming or even expand your live video production.

Capture is the **second "C"** in the business of live streaming. Have you ever been awakened in the early morning with a creative idea? Our first instinct is to shake off the sleep and capture the idea on a piece of paper or mobile device "notes". In reserved quiet time, please take the time to formally document the idea and determine the best way to share the information. This content must be shared with your audience.

As content creators, we deliver engaging and captivating content that will inspire audiences and increase traffic. Capturing your content is an essential part of the content marketing plan. The focus of this book is using Facebook Live to capture and share content. In this example, the content marketing plan will include an outline of the content, target audience, frequency of live streams, schedule of publishing content, and social media marketing.

Live video streaming allows individuals to capture and share your content real time with target audiences. Video is one of the most powerful marketing tools to share content. Creative leaders understand the positive impact that high quality and well produced videos can make on the audience.

If you do not have a Facebook Business Page, the initial goal is to launch your Facebook page to target your primary audience. In the content planning phase, it will be determined what information will be shared via live stream video webcasts. What is the business objective(s)? What are the marketing goals? How to measure success? Is there a call-to-action for the audience? Is the content informational only? Is the content inspirational only?

Multiple Uses of Video (Post Live)

After the live stream has ended, the video is now available on-demand 24 hours a day/7 days a week online. Using video to capture your content has a few advantages, the content can always be reviewed for key information, re-purposed, re-cycled, and re-distributed across other social media sites such as Instagram, YouTube, or Twitter.

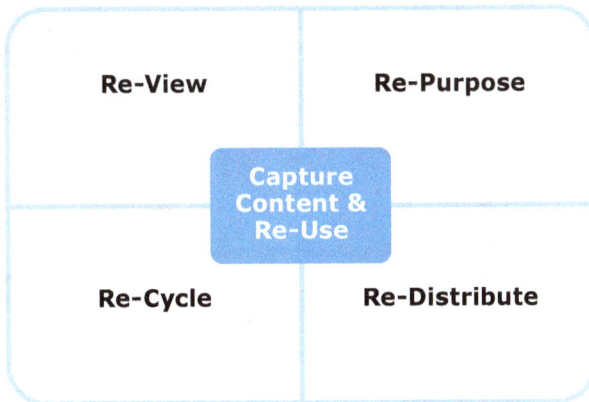

Re-View	Re-Purpose
Capture Content & Re-Use	
Re-Cycle	Re-Distribute

TYPES OF CONTENT TO INCREASE ENGAGEMENT & TRAFFIC

	CONTENT	BIZ TIDBITS
1	Videos	Make your video stand out and add Titles, Backgrounds, Overlays and a business logo.
2	Social Media	Build your audience and promote your products and services
3	eBooks	Great lead magnet to grow your email list
4	Podcasts	Portable, easy to use, and convient for your audience
5	Blogs	Use content to inform, inspire, call-to-action, build relationships, and increase your audience
6	Infographics	Visual display of relevant content included as metrics, statistics and analytics, survey results, etc.
7	Guides, Checklists	These resourceful documents will give the audience to put into practice what you are teaching during your live stream.

BIZ TREND

Eighty percent of people would rather watch a live video from a brand than read a blog or social post. Forrester Research sums up the power of video for engagement: one minute of video is equal to 1.8 million words to your audience.

SOURCE: HTTPS://WWW.STATEOFDIGITALPUBLISHING. COM/INSIGHTS/LIVE-STREAMING-VIDEO-TRENDS/

TIERS OF LIVE STREAMING
BEGINNER OR PROFESSIONAL

The two tiers discussed in this book are for beginners and those who are ready to advance to professional looking broadcasts. There is also a third tier for live streaming for larger organizations, broadcast companies, and major events. These larger organizations or events require more advanced equipment and technology such as media servers, bandwidth, media staff, technical experts, live streaming software that allows for many inputs and outputs for video, audio, and other equipment.

Your financial budget and relevance to your business goals will determine which tier to start or upgrade.

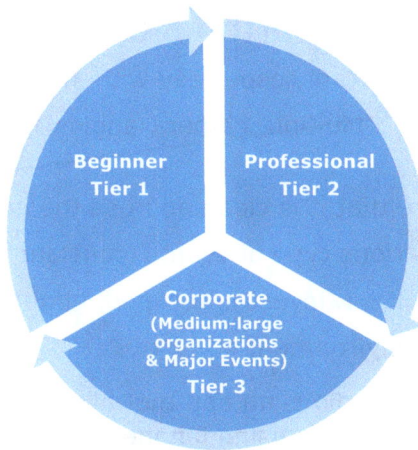

TIER 1 - LIVE STREAMING: Getting Started

Businesses may get started using a smartphone such as iPhone and Android for Facebook Live broadcasts. Using a smartphone is an inexpensive option to get started, this is a digital device that we already use daily. It is better to start now than never. The ultimate goal is to advance to using professional gear and technology.

Here are a few things to consider prior using Smartphone(s):

1. Storage is important, especially with 4K video
2. TYPES OF SHOTS: Multiple camera angles for a more versatile production
3. AUDIO: Ensure that it is clean and use the correct microphone
4. Lenses for various type of events (consider wide angle lenses)
• CAMERA STABILITY: invest in a tripod, gimbal
5. Lighting the environment

Resource: Switcher Studio is an app that allows content creators to film multiple camera angles using iPhone and Android.
Visit www.SwitcherStudio.com for prices and free trial.

TIER 1: LIVE STREAMING : GETTING STARTED

☐	Facebook Business Page
☐	Camera: Smartphones (iPhone or Android)
☐	Tripod / Smartphone mount
☐	Audio: Lavalier Microphone
☐	Lights: Fluorescent or LED Lights
☐	Video Storage for smartphone (get plenty of space)
☐	Editing App: • iMovie • LumaFusion

OPTIONAL

☐	Android only: Oratory (Teleprompter app)
☐	iPhone & Android: BIGVU (Teleprompter app) Padcaster Parrot Portable Smartphone Teleprompter
☐	Wide Angle HD Camera Lens

TIER 2 - LIVE STREAMING: PROFESSIONAL

The professional tier two is for businesses who are ready to produce high quality and professional looking live stream broadcasts. Businesses will need to invest in live streaming production software, train volunteers, or hire crew, and purchase other gear and technology needed to provide a new level of content to their clients.

In the pre-production phase and budget planning, please consider if the business wants to launch single camera or multi-camera live stream broadcasts. In this book I do recommend appropriate gear and technology. (As a Live Stream Consultant, I am available for online and consultation calls for more detailed guidance.)

TIER 2 : PROFESSIONAL LIVE STREAMING

Facebook Business Page	New or existing page
Live Video Streaming Production - Software options	
Wirecast	www.telestream.net
vMix	www.vMix.com
OBS Studio	www.obsproject.com
Camera/Camcorders:	Canon Sony Panasonic
Full size Camera Tripod	Amazon.com
Video Mixer *multi-cameras*	Roland Professional A/V V-1HD HD Video Switcher
Capture Card	AJA U-Tap or Magewell HDMI or SDI
Audio	Lavalier, Handheld, or Wireless Microphones Amazon.com
Audio Mixer	Guitar Center.com, Amazon.com
LED Lighting or Fluorescent	Amazon.com
Laptop or Desktop Computer	Visit Telestream.net/wirecast or VMix.com for system required specifications to run their live video streaming software on your computer
High Speed Internet	Speed requirement will vary but minimum 10mbps and higher upload speed for high definition (HD) video
External Hard drive Video storage and edited files	Amazon.com Bhphotovideo.com

TIERS 1 - 2: TEST INTERNET SPEED

www.SpeedTest.net

As you know, Wi-Fi access can be unstable. I highly recommend that you test and retest your internet connection. For laptop and desktop computers, please use an Ethernet connection (plug-in) to ensure internet connection is reliable.

Please test your internet connection the day before and again on the same day of your live stream. Internet speed does matter. A minimum of 10 mbps upload speed and above for live streaming high definition (HD) video will help to deliver a stable live feed. This will provide the ultimate live stream experience for your audience.

- **UPLOAD SPEED** is the time it takes to send data out, such as your live stream, photo, email, etc. to the internet.

If you are having issues with your internet connection, please contact your internet service provider (ISP).

A few tips:

- Use an Ethernet internet cable (wired) for your live broadcasts. It will provide a stable internet connection.
- Follow-up with your Internet Service Provider to check your upload/download speeds.

3 PHASES OF LIVE VIDEO STREAMING PRODUCTION

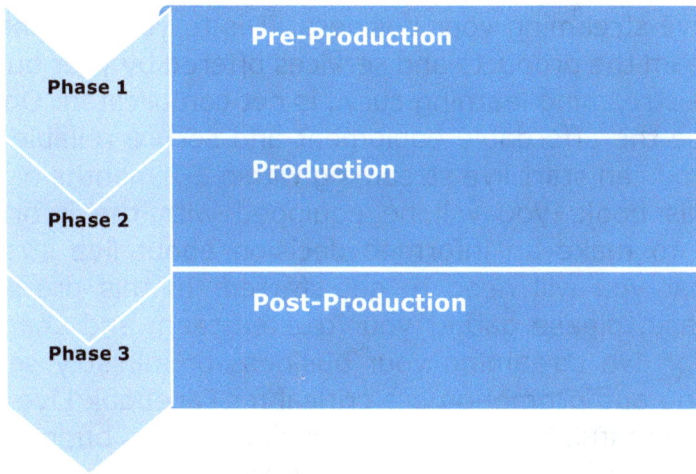

Phase 1	Pre-Production
Phase 2	Production
Phase 3	Post-Production

Once you have determined which tier is best for your business, it's time to get started with live streaming. With live video stream production, there are 3 phases of producing your video content. These phases range from pre-planning your launch through archiving your videos. Following these three phases will definitely help you communicate the broadcast plans with the team and the outcomes expected for each live stream broadcast.

Using these three phases of live stream video production, businesses will have a roadmap to follow for each live stream broadcast, especially for promoting products, services, and events.

PHASE 1: PRE-PRODUCTION

The pre-production of live streaming is the planning phase. Why do you want to live stream? Do you have a budget to get started live streaming? Do you envision your business using live streaming as another source of revenue? These are just a few questions that will need to be answered in this phase of live streaming. This is your planning phase. I want to help alleviate the stress of thinking live streaming will be another big investment that requires you to shell out lots of cash. You will not go broke live streaming your business. It is to your advantage to live stream the products and services offered by your business. The live streaming learning curve is not complicated. Once you purchase the affordable equipment and secure reliable volunteers, you can start live streaming within 1-2 months or less.

In this book, you will be equipped with the information needed to make an informed decision about live streaming and what you will need to get started. In this pre-production phase, please gather your trusted team and share your vision for live streaming your business or ministry services. Marketing and promotion are critical for Facebook Live events to be successful. If you do not have the financial budget to hire a social media manager please ask for a volunteer on social media or contact your local community colleges or universities to recruit an intern to manage your Facebook business page, to set up Facebook advertisements, and other social media sites. The social media intern will need the work experience for their career advancement, and they would love to be on your team.

BIZ TIDBIT

Without a plan, the vision will fail.

SHERRY BRONSON

PRE-PRODUCTION LIVE STREAMING
THINGS TO CONSIDER

1. Why do you want to Live Stream?

_____

_____

_____

2. What content do you want to share?

_____

_____

_____

3. How do you plan to accomplish it?

_____

_____

_____

4. When do you plan to begin?

_____

_____

_____

5. What social media platform(s) will you use to share your content?

_____

_____

_____

6. Will you have the budget to hire a social media manager to manage your content on social media platforms?

_____

_____

_____

Note: Consider recruiting volunteers until you can afford to pay for staff.

USE VIDEO STORYBOARDING

Storyboarding is very useful when there is a need to provide a visual representation to the staff and media team prior to going live. Storyboarding is an optional task in pre-production planning. It is an outline of the video shots, angles, and set design that will be captured during the broadcast. It helps to map out the entire live stream and gives you an opportunity to place it on paper as a reference point for the production crew.

Storyboarding will help with explaining ideas and vision for the live stream production. Storyboarding is not required. Hiring a Storyboard artist to draw is not necessary.

Please use a free downloadable Microsoft Word storyboard template, please print the template, write out the details of what is envisioned, and make it simple.

Storyboarding can be used for the very first Facebook Live stream only. As the media team gains experience, please make necessary edits to the original storyboard and then follow the same live stream format.

Video Storyboarding Template Example

Video Storyboard

Equipment Needs: Check out date_____ Return Date: _____

☐ Camera _____ ☐ Audio _____ ☐ Other _____

☐ Tripod _____ ☐ Tape _____ ☐ Other _____

A storyboard is probably the most useful planning document for both inexperienced and professional video makers. It consists of frame-by-frame sketches that indicate the content of each scene and the way in which it will be shot—that is, shot type, angles, and so on. Accompanying notes can be used to describe what you are trying to achieve in each shot as well as audio notes. You do not have to be a great artist to produce a useful storyboard since it is generally only for your own reference. The advantage of this kind of document is that it helps you to visualize not only individual scenes but also the overall pace of the action and atmosphere of your planned video.

Production Title: _____

	Scene Sketch	Shot Description	Audio	Comments
1				Time:_____
2				Time:_____

Production Title: _____

Page 2   Home Education Exchange

Resources: Downloadable Word Templates

- StudioBinder.com
- Storyboardthat.com (recommended)
- Template.net

PRE-PRODUCTION PLANNING SESSION	
Topics of Discussion: Facebook Live Webcasts	
Budget	Purchase gear, tools needed
Content	What information to share? How often?
Live Production Schedule	Write a schedule draft and share with the entire crew
Location	Where to live stream? Studio design
Crew	Hire videographer or use volunteers
Training	Train your crew member(s): Who will conduct training?
Equipment:	Buy or Rent
Single Camera	1 video camera
Multi-camera production	2–3 Cameras: How many video cameras are needed?
Audio	Audio mixer/microphones
Video Mixer	Use for Multi-camera production
Lighting	Lighting technician to assist
Internet	High Speed Internet / Ethernet Connection
Social Media	What other social media sites to distribute content?
Archiving Content	External hard drive, cloud
Re-distribution	Re-Cycle and Re-purpose live videos

PHASE 2: PRODUCTION

Let the Filming Begin! Lights, Camera, & Action

The production phase is the actual live video webcast on Facebook. The media team is now ready to retest the internet connection and equipment for the live stream. Your media team will need to do a final review and make last minute edits to your live stream broadcast schedule prior to distribution to your leaders and team members.

The production phase is the same day of your Facebook live event. Please have the media team at the live event location early for preparation and instruction. Marketing will be very essential to the success of your Facebook Live webcasts. The social media manager or volunteer will share in advance and on the same day the date, time, and Facebook page location of your live webcast.

It is beneficial to use Facebook advertisements (ads) to promote the Facebook Live webcast. Facebook advertisements allow us to target our audience and make them aware of the live events. Facebook ads allow users to set their daily budget without unexpected costs.

BIZ TIDBIT

An unexecuted vision is just a daydream.

GO LIVE

Live Broadcast Schedule
Example 1

Live Video Stream Broad Schedule Draft: Example 1		
Webcast #	**Topic**	**Duration**
Webcast 1	Get to Know Us: Q&A	1 hour
Webcast 2	Tips & Free Give-away	1 hour
Webcast 3	New Book Launch	1 hour
Webcast 4	New Product Demo	1 hour
Webcast 5	Webinar: How to?	1 hour
Webcast 6	Panel of Business Experts	1 hour

Example 2

Live Video Stream Broadcast Schedule: Example 2					
Webcast #	**Topic**	**Date**	**DURATION**	**Start Time**	**Crew Name**
Webcast 1					
Webcast 2					
Webcast 3					

Please add additional columns to the live production schedule as needed. For effective communication, please share the production schedule with the entire media team.

PHASE 3: POST-PRODUCTION

Once the Facebook Live event has ended, please have a plan to re-distribute the video content. Post-production does require editing the video content for re-distribution. Video editing is essential to provide a more professional look to videos. Editing features include adding graphics such as lower thirds (titles), high quality images, royalty free music, visual effects, sound effects, and more. Hiring a video editor can be costly but learning to edit your videos will save lots of money. If the live webcasts are 30 minutes to 1 hour, please select the best clips and edit them to 1, 2, or 3-minute video teasers and soundbites. Please share the edited video teasers and reels on social media and embed on your website. With the short attention span of internet users, videos have to have a hook to grab their attention. We want to build an engaged audience and that is possible with great substantive content and high-quality videos.

Filmora video editor software has an easy learning curve and it is affordable. It does allow users to produce great looking videos. I do recommend this video editing software. Please watch their video tutorials and get started.

Visit www.wondershare.net for Filmora software pricing.

BIZ TIDBIT

Great content never gets old, share it and share it again.

4 Popular Live Video Streaming Production Software Options

There are many options available for live video streaming production software. These are some of the most reliable options for budget conscious business owners, houses of worship, speakers, trainers, schools, and other professional organizations.

Please visit each vendor's website for details. Prior to purchasing the live stream software, the vendors offer free trials to test their software. I highly recommend that you test it with the software to ensure that the live stream will be seamless.

Live Video Streaming		
Production Software	Website	Costs
1. Wirecast	Telestream.net	$249 & above
2. vMix	vMix.com	$60 & above
3. Open Broadcaster Software (OBS)	Obsproject.com	Free
4. NewTek Tricaster	Newtek.com	Contact Vendor

Live Video Streaming Production Software
Option #1

Wirecast

Telestream.net software

Prior to making a decision to purchase live streaming production software, it will require research and testing. It is imperative that the right software is chosen to accomplish the business needs. This is an important financial investment.

The live video streaming production software is fantastic for taking videos to another level. The features are multiple inputs and outputs, social media comments, b-roll videos, media files to be imported such as audio, video, images, pdf files, and many other features to create professional looking live webcasts.

I have been a happy Wirecast customer for five years. It has really been a great experience using Wirecast's live video streaming production software. They are constantly making improvements to their platform. We are currently using Wirecast Pro version and it is easy to navigate the user interface.

Visit www.Telestream.net to compare their live streaming options at various versions and prices

Live Video Streaming Production Software
Option #2

vMix

Vmix is another live video production streaming software that is popular among businesses, ministries, and other content creators. The price is very affordable, and they offer multiple software options to purchase from beginners to professional. We use vMix for live webcasts. They have great features such as vMix Call and vMix Social software, test out their free version. Visit www.VMix.com for pricing, additional features and more details

LIVE VIDEO STREAMING PRODUCTION SOFTWARE
Option #3

Open Broadcaster Software

Open Broadcaster Software (OBS) is a free and open source software for video recording and live streaming. If there are budget restraints, OBS can be a first option. The media team will be able learn the software to get started. Go live on Facebook!

Here are a few of the OBS features:

1. Add lower thirds (Titles)
2. Add Media files (Logo, images, audio, video, pdf)
3. Custom background using Greenscreen/Chroma key

Visit http://obsproject.com to download and details

Live Video Streaming Production Software
Option #4

NewTek is an advanced video production system and software available for medium to large organizations, houses of worship, and major events. This will require a large financial investment for equipment, bandwidth, additional knowledgeable crew, and technical support.

Visit www.Newtek.com for a demonstration and costs

GEAR: STARTER KIT

SINGLE CAMERA LIVE PRODUCTION

For small businesses and houses of worship with small budgets, please feel free to start live streaming with a single camera. There is no pressure to go out and purchase expensive gear to begin live streaming. The single most important piece of gear that you own right now is a Smartphone. We all know that the smartphone is a minicomputer that connects us to the world. Many smartphones have a high-quality camera and built-in microphone that allows us to download the Facebook app and other social media apps. Please purchase a smartphone mount holder and a desktop or full-size tripod to avoid shaky looking videos. Once the Facebook business page has been designed and published, it is time to begin live streaming.

Mobile Video Editing app for iOS/Apple & Android

Please keep in mind there will be postproduction video editing needed. Live webcast can be one hour long or longer. Once the live streams have ended, the video becomes available as on-demand. I bet during the full live webcast there are some captivating sound bites and key information that needs to be re-shared with everyone. The great news is that those videos can be re-purposed by using video editing software to trim them down to 2 to 5 minutes. Once the videos have been trimmed down, please create new thumbnails for each video, a catchy video title, and schedule them to auto post over a day, weeks, or even months. There will be plenty of on-demand (post-live) videos that will need to be re-shared and re-purposed.

Here are a few mobile apps for video editing for smartphones: iOS or Android

Mobile video editing apps features include trim video, add titles, add music, adjust color, effects and more. Once editing is completed, please share on Facebook and other social media platforms.

1. iMovie

2. LumaFusion

3. Videoshop.net

4. InShot

5. KineMaster (Android & iPhone)

6. FilmoraGo (Android)

GEAR PACKAGE #1

GETTING STARTED		
Single Camera Live Streaming	**Vendor**	**Approximate costs**
Smartphones / Mobile Devices: iOS / Android phones iPads	Apple Samsung	Already a part of your toolkit
Option 1: MeFOTO SideKick360 Smartphone Tripod Adapter	BhPhotoVideo. com Amazon.com	$32.95
Option 2: JOBY - GorillaPod 3K SMART Kit Tripod	Bestbuy.com	$59.99
Full Size Tripod Brands: Amazon Basics Neewer Manfrotto GEEKOTO	Amazon.com	$25.99 and above
Audio Option 1: BOYA by M1 Lavalier Microphone for Smartphones	Amazon.com	$19.95
Audio Option 2: Rode VideoMicro Compact On-Camera Microphone Kit for Smart phones	Bhphotovideo. com	$74.00
Increase Video Storage for recording and archiving	Smartphone provider	Pricing will vary. Contact your smartphone service provider

GETTING STARTED ON A SHOESTRING BUDGET WITH SMARTPHONES

If you are planning to get started or continue to use smartphones, iPads, etc. to broadcast live, please visit www.iOgrapher.com. iOgrapher provides affordable gear bundles and accessories for live video streaming. iOgrapher is also offering interest free payments using sezzle.

Live Video Starter Bundle for Tablets

- iOgrapher Flexible Tripod
- iOgrapher Mini Ring Light

Starter Live Video Bundle for smartphones

- iOgrapher Tablet Holder
- iOgrapher Flexible Tripod
- iOgrapher Mini Ring Light
- iOgrapher Ring Light for smartphones
- iOgrapher Ring Light comes with a tripod stand and cell phone holder.
- USB Charging Cable
- Wire Controller Remote
- Light and Phone Holder
- Tripod

GEAR PACKAGE #2

The purchase of the equipment and technology will depend on your budget.

CAMERA CAMCORDER		
Single Camera Production	**Vendor**	**Approximate costs/notes**
Laptop or Desktop Computer	Mac (Apple) Windows	Upgrade or New PC to meet the system specification requirements for the live streaming software
OPTION 1: Consumer Camcorder Canon HF R800 Bundle HD Recording Portable Traditional Video Camera, Black Panasonic Full HD Video Camera Camcorder HC-V770 OPTION 2: PTZ Optics USB or HDMI connection	Amazon.com PTZOptics.com Bhphotovideo.com	Canon HF R800 $260 & above Panasonic HC-V770 $599 (research other models and brands Canon, Sony or Panasonic)
OPTION 3: DSLR or Mirrorless Camera	PTZOptics.com Bhphotovideo.com	Affordable and prices vary
AJA Video Capture Card Blackmagic Design Ultrastudio Mini Recorder Elgato Cam Link 4K	Bhphotovideo.com	1. $345.00 2. $145.00 3. $129.99
OPTION 1: Behringer UM2 Audio Interface for audio capture (USB) (1 channel) OPTION 2: Behringer Xenyx 302USB Premium 5-Input Mixer with Mic Preamp and USB/Audio Interface, Black	Amazon.com	Option 1: $48 Option 2: $78.99
LIVE VIDEO STREAMING SOFTWARE (ENCODING SOFTWARE) Wirecast (telestream.net) vMix.com obsproject.com (free)		Visit their website for prices and free trials
LED Lights	Amazon.com Bhphotovideo.com	Brand prices will vary
FULL SIZE TRIPOD BRANDS: Amazon Basics Neewer Manfrotto	Amazon.com	$25.99 and above
ACCESSORIES: Memory card (SDHC or SDXC) HDMI or SDI cable	Amazon.com Bhphotovideo.com	
EXTERNAL HARD DRIVE: Archive videos, Edited videos	Amazon.com Bhphotovideo.com	Prices will vary- westerndigital.com glyphtech.com www.lacie.com

GEAR: PROFESSIONAL

MULTI-CAMERA LIVE PRODUCTION

As your live streaming experiences increase, begin to evaluate if it is the right time to invest in more professional gear, such as live video streaming software, multiple cameras, video switcher and more. With additional professional gear, it expands your workflow and creative genius to produce amazing eye-catching videos.

For post-production, please purchase or Upgrade Laptop or Desktop computer. The computer's performance will definitely need to be powerful to handle the live stream to multiple platforms such as YouTube and Facebook. The internet speed will need to be faster, especially upload speed.

CAMERA CAMCORDER		
Single Camera Production	Vendor	Approximate costs/ notes
Laptop or Desktop Computer	Mac (Apple) Windows	Upgrade or New PC to meet the system specification requirements for the live streaming software
OPTION 1: Consumer Camcorder Canon HF R800 Bundle HD Recording Portable Traditional Video Camera, Black Panasonic Full HD Video Camera Camcorder HC-V770 OPTION 2: PTZ Optics USB or HDMI connection	Amazon.com PTZOptics.com Bhphotovideo.com	Canon HF R800 $260 & above Panasonic HC-V770 $599 (research other models and brands Canon, Sony or Panasonic)
OPTION 3: DSLR or Mirrorless Camera	PTZOptics.com Bhphotovideo.com	Affordable and prices vary
AJA Video Capture Card Blackmagic Design Ultrastudio Mini Recorder Elgato Cam Link 4K	Bhphotovideo.com	1. $345.00 2. $145.00 3. $129.99
Behringer Xenyx 302USB Premium 5-Input Mixer	Amazon.com Bhphotovideo.com	$78.99
Live Video Streaming Software (encoding software)	Wirecast (telestream.net) vMix.com obsproject.com (free)	Visit their website for prices and free trials
LED Lights	Amazon.com Bhphotovideo.com	Brand prices will vary
FULL SIZE TRIPOD BRANDS: Amazon Basics Neewer Manfrotto	Amazon.com	$25.99 and above
ACCESSORIES: Memory card (SDHC or SDXC) HDMI or SDI cable	Amazon.com Bhphotovideo.com	
EXTERNAL HARD DRIVE: Archive videos, Edited videos	Amazon.com Bhphotovideo.com	Prices will vary westerndigital.com glyphtech.com www.lacie.com

10 CREATIVE WAYS TO RE-PURPOSE YOUR VIDEOS

- **DIGITIZE VIDEO:** Convert videos to MP4 (video) and MP3 (audio) files and monetize them
- Create multiple 1-2-minute bite-sized video clips of key moments of the video
- Post 60 second Instagram videos
- Post to Twitter Tweets (2 minutes 20 seconds)
- Upload videos to your YouTube Channel
- Post informative videos on LinkedIn
- **VIDEO TO TEXT:** Get your video/audio transcribed to text, add the text to your videos post to your social media sites
- **VIDEO TO TEXT:** Create an eBook
- **BLOG:** Grab high quality screenshots of your videos and write a blog about a helpful topic
- **QUOTES:** Use quotes from the videos, add to video screen grabs from the video and post to your social media sites

Resources:

- Snagit.com (screen grab)
- Canva.com
- Rev.com (transcribing video/audio to text)

CHAPTER 4

CONVERTING LEADS
CUSTOMER BUYING PROCESS

Here are six steps of the customer buying process that we must always keep in mind. This process should be incorporated in your marketing strategy to help with lead conversion, customer satisfaction, and customer loyalty. Convert is the **third "C"** in the business of live streaming. Customers remain loyal when they are valued by businesses, especially if your product or service make their lives better. Your business should be able to address each of these steps.

1. **PROBLEM OR NEED:** Customer has a problem or need. Example: video production, computer, car, home, or health insurance, make-up, installation or repair, software

2. **RESEARCH SOLUTIONS:** Customer begins researching solutions on the internet, Google, social media, asking friends and family, etc.

3. **SOLUTION EVALUATION:** Customer will contemplate, evaluate, or test solutions prior to making a final decision

4. **DECIDE:** Customer has determined the best product or service that resolves their problem or need. Small businesses can help with the customer decision by having a simple process for buying your product or service.

5. **MAKE PURCHASE:** Customer makes the purchase. How was their purchasing experience? Did they experience great customer service?

6. **CUSTOMER SATISFACTION:** Now the purchase has been completed. The customer can change their mind about the purchase. What customer incentives do you have in place to retain customers? How will you prevent buyer's remorse?

UTILIZE FACEBOOK AS A LEAD GENERATOR

Have you ever struggled with increasing traffic to convert a potential lead? If yes, you are not alone in this effort. Finding and converting potential leads into a paying customer is not easy. When we are promoting our services and products, yes, one of our goals should be to make a sell. However, we must make an effort to build a relationship with potential leads and continue to nurture current customers. The advantage of Facebook is the access to over 2 billion active monthly users globally. With this type of statistic, small businesses should use Facebook Live as a lead generator. Businesses have a great opportunity to target their ideal customers using Facebook advertisements. Facebook allows small business owners to drive up enough traffic to be able to qualify leads. The platform allows businesses real-time engagement with potential leads. Small businesses are missing a grand opportunity if they do not have a presence on Facebook or any other social media platform. Facebook is the new

DEFINITION

What is a lead? A lead is any person who indicates interest in a company's product or service in some way, shape, or form.

HUBSPOT.COM

digital word of mouth for businesses. Through Facebook, we are now connected locally, nationally, and internationally. Let's take advantage of this powerful platform!

I encourage you not to feel overwhelmed when contemplating using Facebook to generate potential leads. We understand the many roles that small business owners have to play when building their businesses. When operating your business on a limited financial budget, the roles and responsibilities will shift between sales to bookkeeping, social media manager, and more.

The daily grind to generate new leads is hard work. It is necessary to tap into your inner sales and marketing abilities or find someone with this expertise to help your business.

Have you ever wondered about the process of converting your lead to a paying customer? We have been told to convert potential leads, read about marketing, sales, and lead funnels to help with generating potential leads and drill down to find our ideal paying customer.

DEFINITION: WHAT IS LEAD GENERATION?

Lead generation describes the marketing process of stimulating and capturing interest in a product or service for the purpose of developing a sales pipeline.

MARKETO.COM

I wanted to share some information that I have learned and how to avoid the mistakes that I made. When I launched Bronson Media, I was in this overhyped start-up mentality. There was just so much excitement in my heart and I was overcome with joy to be living out my divine purpose. I did not understand the power of analytics and researching to find your target ideal customer. I discovered that business growth comes from being strategy-driven, not just being driven by happy emotions. Of course entrepreneurs must have a passion about what they do. Yes, people love our enthusiasm, but they care more about us solving the problems that they have.

The information that I share about the process for customers buying products and services will be very helpful. The type of customers' problems or needs can vary from educational tools, software, counseling, etc. In order to reach these customers, it is essential to have a marketing plan that targets your ideal customers. Potential leads are strolling through sponsored advertisement daily in their Facebook news feed and other social media platforms from other competitors.

Utilizing Facebook Live as a lead generator to drive traffic and awareness of your great business products or services is a winning strategy. Potential leads are seeking value for their hard-earned money.

BIZ TIDBIT

Your leads have to first qualify and then they are more likely to convert to buyers.

SOURCE: SALESFORCE.COM

USE LEAD GENERATION TO INCREASE SALES

LEAD GENERATION MARKETING CAMPAIGN	
Use Landing pages for selling products, services	
1	Business has a need for awareness of personal brand, services, and products being offered
2	**TRAFFIC:** Ideal Customers on Facebook
3	**LANDING PAGE BUILDERS:** Free Trial & Paid Plans GetResponse.com, Unbounce.com, or Leadpages.net
4	Video to promote the product or service that solves your ideal customer's problem (1:00-2:30 minutes)
5	**BUILD A CUSTOM LANDING PAGE:** 1. add Video promo 2. Create a Lead Magnet: Example 1: Discount on webinar or course Example 2: Free download (eBook, Handout, etc.) Launch your Landing Page: Create a post on your Facebook Business Page Facebook Advertisement
6	**LEAD CAPTURE OPT-IN:** Lead provides their name, email address and/or other contact information after viewing your promotion
7	Include a Thank you page
8	Follow-up with qualifying leads

Also, visit Facebook.com to watch their how-to tutorial video on creating Facebook advertisements to generate traffic.

GENERATING LEADS

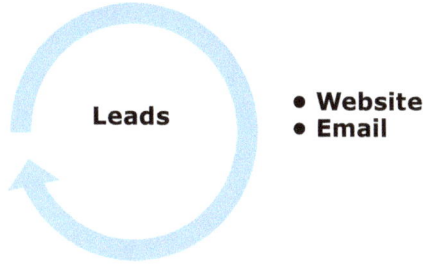

Leads
- Website
- Email

1. WEBSITE: OPTIMIZATION

- Add a **POP-UP FORM** for website visitors to provide their full name and email address and offer a valuable free resource that they can download to their computer. Ask your website designer to make this change.

- **FREE OR SPECIAL OFFERS** can include an eBook, checklist, cheat sheets, workbook, or guide.

POP-UP FORM EXAMPLE 1: TIP SHEET

POP-UP FORM EXAMPLE 2: SPECIAL OFFER

2. EMAIL

Start or continue building your email list through Facebook, Instagram, etc. Create online marketing campaigns to offer rewards for registering for webinars, etc.

Resources: MailChimp.com, MailerLite.com or Constantcontact.com

Example 1: Checklist

Example 2: Free Webinar

CHAPTER 5

CASH: MONETIZE YOUR CONTENT

In this information age, many businesses are leveraging social media to make money. Cash is the **fourth "C"** in the business of live streaming. In order to generate revenue, you must have an engaged audience. There are many ways to make money online. However, I had you in mind when I decided to write this book to share the business of live streaming. How can small businesses make additional revenue using live streaming? There are trainers, coaches, speakers, educators, communicators, and other influencers grinding every day to generate revenue to pay operational costs and hire virtual assistants, etc.

As content creators, we have to always think outside the box and develop new content, concepts, and strategies to expand our vision. Currently, there are many small businesses that are using Facebook, Instagram, and other social media platforms to monetize their content and generate revenue. Please take into consideration that it is not easy to build a diverse and loyal audience on social media. It will be gradual growth, but you will grow your audience with hard work. From experience, it requires us to have consistent engagement with our new and existing audience. We have to be very strategic and relentless about sharing new, re-purposed, and re-cycled content daily or three-four days a week.

If you are new to social media, I highly recommend that you choose one social media platform and grow your loyal audience. This is essential before you can launch a product and/or

> **DEFINITION**
>
> Monetize: Earn revenue from (an asset, business, etc.)
>
> DICTIONARY.COM

service to begin selling. Please keep in mind that your engaged audience will make purchases and tell their friends and family about your wonderful brand, products, and services.

5 KEYS STEPS TO MONETIZE YOUR CONTENT

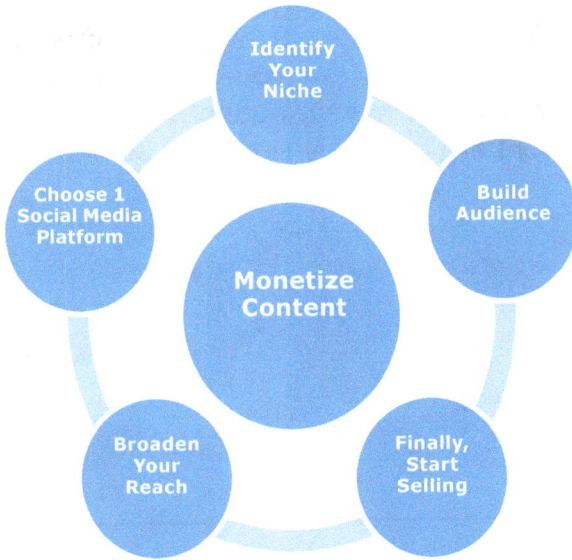

Identify Your Niche

Choose 1 Social Media Platform

Build Audience

Monetize Content

Broaden Your Reach

Finally, Start Selling

HOW TO MONETIZE YOUR CONTENT?		
	Key Steps	Biz Tidbits
1	Identify Your Niche	What's your area of expertise? Who is your intended audience? What products and services do you want to sell?
2	Choose one Social Media Platform	Grow your base on Facebook Business Fan Page
3	Build Your Audience	If you are new to social media, audience growth will be organic, be consistent with posting regularly and also accelerate growth through Facebook ads
4	Broaden Your Reach	Expand to other social media sites Instagram, Twitter, etc.
5	Launch Your Products and Services	Start selling on social media

Get started monetizing your content and make an assertive effort to build your audience on Facebook as soon as possible.

GENERATE REVENUE WITH YOUR CONTENT

What to do with your video content after Facebook Live? So far in this book, I have shared with you two phases of live streaming. We are now in phase three, Post-production of live streaming. Businesses have plans for their videos after Facebook live webcasts are completed. After you go live on Facebook, many content creators do not re-distribute or re-purpose their videos. Content creators are definitely sitting on some revenue.

WAYS TO GENERATE REVENUE WITH VIDEOS

	CREATE DOWNLOADABLE DIGITAL PRODUCTS	
	Video Content	Biz Tidbits
1	Audio (MP3)	Convert your Video files to MP3 audio files Sell your Audio files as downloadable resources on your website and social media
2	Video (MP4)	Convert your videos to smaller MP4 files Sell your Video files as downloadable resources on your website and social media
3	eBooks	Transcribe audio from videos to text Hire a transcription business to convert your video to text; Sell eBooks on your website and social media Resources: www.Rev.com or Fiverr.com
4	Workbook	Schedule a Free Weekly Live Webcast Series (4 weeks) to teach on a resourceful topic Transcribe the video to text Schedule a Paid Webinar or Online Course and include a workbook or handbook as part of the participant's registration fee Resources: Udemy.com or Teachable.com

OTHER WAYS TO GENERATE REVENUE WITH VIDEOS
Live Pay-Per-View (PPV) Event

There are various companies that offer pay-per-view services for live events and conferences. These types of live events will allow you to broadcast live and sell your videos live and on-demand. In this digital age, we must take advantage of these powerful software platforms. I am recommending a few software companies that are budget friendly and they will walk you through the process. This is an opportunity to generate additional revenue through pay-per-view events and selling your content live, on-demand, or subscription packages.

Please contact each business to schedule a demonstration to ensure that you understand the process of setting up the online video player (OVP) and pay-per-view live events. I have provided some things to consider prior to contacting the service provider sales team. This will help you with determining your budget for the live pay-per-view event. You want your live PPV events to be profitable for your business to invest in marketing and hosting costs for the live events.

LIVE PAY-PER-VIEW (PPV) EVENTS

There are 2 key software providers that you will need to contact prior to hosting your live Pay-Per-View events. I have listed below a few professional companies for you to contact for price quotes and demonstrations.

1. **ONLINE VIDEO PLAYER:** Video Player Software Company
2. **PAY WALL:** Monetization Company

THE VIDEO PLAYER SOFTWARE manages the live stream. Your participants will be watching the live event from the embedded video on your website or any other social media platform that you choose.

#	Online Video Player: Software companies
1	JWPlayer.com
2	Vimeo.com
3	Brightcove.com

THE PAY WALL will manage the revenue generated from ticket sales, subscriptions, video sales, and other analytics.

#	Pay Wall: Participants purchase their Tickets
1	InPlayer.com
2	Streamingvideoprovider.com
3	Cleeng.com

Live Event	Number of Viewers	Ticket Price	Revenue Event Ticket Sells	Sell Videos	Optional: Subscription

Here are a few things to consider prior to contacting the Video Play Software and Pay Wall companies.

SERVICE PROVIDER PAY WALL – SELL THE VIRTUAL TICKETS	
Content	Live and On-Demand: What type of content do you plan to sell?
Online Video Player	
Event Details	What is your estimated online viewership for your live event? What is the Ticket Price? Replay: What will be the subscription fee for viewing on-demand content?
Website	Where to watch the Live Event? Will the video player be embedded in your website?
Selling Virtual Tickets	What percentage does the service provider receive from each virtual ticket sold?
Selling Videos	Do you get 100% of the profits?
Hosting Costs	Monthly or Yearly costs What is the cancellation policy?
Bandwidth	What are the additional bandwidth rates especially if you expect to register over 100 participants?
Video Recording	How to access videos after the Live Event?
Support Plan	Are they available via phone during your first live stream? What's included in their support plan?

CHAPTER 6

FACEBOOK LIVE FOR SMALL BUSINESSES

Getting Started

Facebook Live launched on April 6, 2016. It is a powerful social media platform for businesses worldwide. Social media helps connect online globally in real time. Using Facebook, businesses can build an online community to find new customers, educate, and inspire viewers. As we are experiencing, social media has definitely made the world much smaller.

The digital age has descended upon us. We must accept this change, adapt, and implement digital media into our business practices and marketing strategies. Fortunately, Facebook is just one social media platform that you can use to begin to learn and gain experience with building an audience. One of the keys to success with using Facebook Live is there must be a plan of action.

PUBLISH A FACEBOOK BUSINESS PAGE

The success or failure of a Facebook Live Business Page launch is all in the hands of the business owner. The first step is to complete all the appropriate areas on the Facebook business page and then publish it to the world. As you are editing the new Facebook business page, it will remain in an unpublished status. Please do not feel rushed to publish the page. Set your launch date and work hard to meet it and go live. Below are a few pre-launch tips. The goal is to begin to build your audience as quickly as possible. Build your momentum in the first 90 days. Stay active and consistent.

FACEBOOK PAGE GROWTH TIPS

1. FACILITATE A PRE-GO LIVE TEAM MEETING:

- Please include your social media manager and graphics designer. Topics of discussion will include, who will be the administrator to manage the page, Facebook page design, page cover, and profile picture

- Decide if you will use a mobile device. (iPhone, iPad, or android to go live)

- Decide if you will invest in live video streaming production software for a more professional live stream experience for your audience.

2. FACEBOOK ADVERTISEMENT & BOOSTING

- Facebook allows businesses to create ads and boost posts to promote their pages, products, services, and to grow their fan base. It will require a budget friendly investment from the business owner. Here is the difference between an ad and a boost in Facebook.

- **FACEBOOK ADVERTISEMENT:** "While boosting a post is still considered an ad, Facebook ads are created through Ads Manager and offer more advanced customization solutions. There are many advertising objectives to help you reach your specific business goals and the audiences you care about most."

- **BOOST A POST:** "If you want audience engagement on your Page or to develop your brand awareness, boosting a post is a great way to maximize visibility and grow your audience. To create more advanced ad types and campaigns, use Ads Manager." (Source: Facebook.com/Business)

3. CANVA

- If you are a solopreneur, I highly recommend making a small investment into a graphic design tool called "CANVA" to create your graphics. (www.Canva.com)

FACEBOOK LIVE: MARKETING PLAN USING S.M.A.R.T GOALS

To avoid wasting time and money, it is important to have marketing goals that are attainable, I recommend using S.M.A.R.T. goals to achieve those goals.

What are S.M.A.R.T goals?

- **SPECIFIC:** Write measurable goals: Define outcomes and deliverables. How Much? How many? When is the goal accomplished?

- **MEASURABLE:** Write measurable goals: How do you measure success? How many? When is the goal accomplished?

- **ATTAINABLE:** Specify who will help achieve the goals? Goals should be attainable, ambitious, and action-oriented

- **REALISTIC:** Is it worthwhile? Does it align with our vision? Is this the right time? Do we have the finances for this endeavor?

- **TIME-BOUND:** Time oriented, time-based/deadline, time/cost, timeframe.

S.M.A.R.T Goal Example

- Increase Facebook Page Likes by 1000 in the next 60 days using Facebook Ads and Boosting Posts

S=Specific	What do I want to accomplish?	Increase Facebook Page Likes by 1000
M=Measurable	How will I measure progress?	View your Facebook analytics and insight reports on a weekly basis to see your progress
A=Attainable	Action-oriented: How can the goal be accomplished?	Social Media Manager will post content 3-4 days per week Business owner will host weekly live webcasts for 60 days
R=Realistic	Is this a worthwhile goal? Do we have the finances to achieve this goal?	Determine if you have the finances and human resources available to accomplish the marketing goal.
T=Time-Bound	How long will it take to accomplish the goal?	Marketing Campaign Duration: 60-days Investment: Total $300 investment in Facebook Ads & Boosts for 60 days) Of course, invest additional monies if needed.

There is a great feature in Facebook that allows users to schedule posts in advance. Users are not under pressure to post daily, but they can reserve a specific day to schedule posts to appear on the page then boost it.

1. CONSISTENT LIVE WEBCASTS

- Promote all your live webcasts in advance to build the excitement.

- Schedule weekly live webcasts as an event and then invite the page fans. Fans will get a reminder of the date and time of your webcasts. Fans can share the event with their friends.

2. VIDEO TAB: CREATE VIDEO PLAYLISTS

- Create video playlists will help customers and followers find relevant content and keep your videos organized.

- **PLAYLIST CATEGORIES:** Live Tutorials, Live Product Demos, Personal Vlogs, Promos, etc.

- Create a **CUSTOM THUMBNAIL** image for videos to make them professional. Image needs to be high quality. Use Canva to create a graphic.

BIZ TIDBIT

Facebook is the mobile solution for 65 million businesses with Pages and 8 million with Profiles on Instagram.

FACEBOOK.COM/BUSINESS

FACEBOOK ANALYTICS: HOW DO YOU DEFINE SUCCESS?

When investing money and time into marketing campaigns, we would love to experience a Return on Investment (ROI). Please do not fall into the trap of comparing your Facebook Business Page likes and engagement to other businesses. Everyone has a different starting point, social media education level, and resources to support and grow their Pages.

When launching Facebook Ads and Boosting posts, please evaluate the Facebook Insights (analytics) results at the end of each campaign. Prior to setting up Facebook ads and boosting posts, please decide on these 5 demographic areas to reach new leads or customers that you are targeting. If you are not satisfied with the results, please make necessary changes to the Ads and re-launch new campaigns.

Demographics

1. Audience
2. Location
3. Gender
4. Age
5. "Interests" Exclude or Include

What do you consider success for your Facebook Business Page marketing campaigns? Setting realistic goals for marketing campaigns will guarantee some success. Ultimately, the small business owner will define the success of the Facebook marketing campaigns. For instance, do you want 1,000 likes each month and $500 product purchases each month? These types of goals will be included in your marketing strategy.

Also, please keep in mind that Facebook has an algorithm that decides which posts users will see on their newsfeeds. With frequent and consistent posts and live webcasts on your Business page, you will have audience engagement.

Facebook mentions three major categories of ranking signals:

1. Who a user typically interacts with

2. The type of media in the post (e.g., video, link, photo, etc.)

3. The popularity of the post

Source: https://blog.hootsuite.com/facebook-algorithm/

FACEBOOK "BeLive Studio"

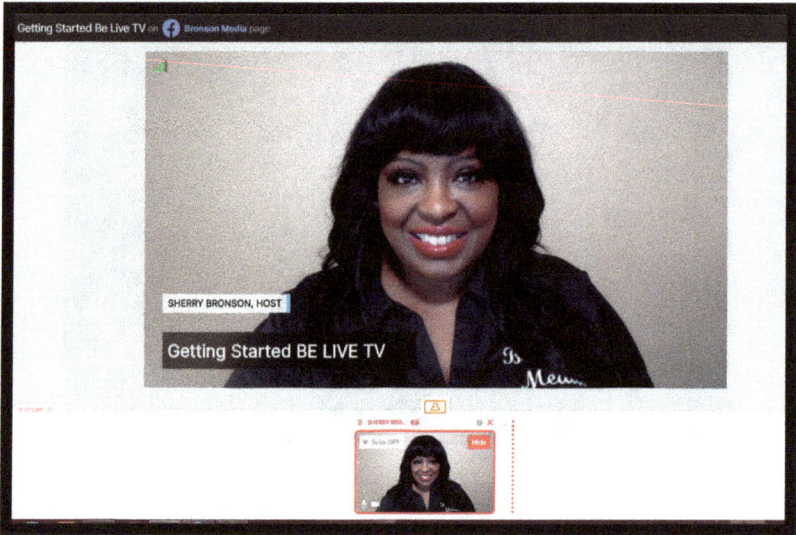

Professional Live Streaming Software:
www.BeLive.TV

For small business owners, communicators and other professionals who are exploring options to start professional live streams. Facebook has a great live streaming application called BeLive TV. BeLive is available for free to gain experience prior to making a financial investment in a more robust live video streaming software.

The app is easy to navigate and customize. The paid subscriptions offer additional features, especially if there will be additional guests or co-hosts. I actually love how easy it is to invite guests. It allows you to change the view modes for a very engaging and interactive live stream experience. Please prep your guests prior to the live stream to test audio and video settings. This will definitely help with delay and annoying technical difficulties. I repeat, test, test, and test in advance.

BELIVE TV

FACEBOOK APP FEATURES:

- They have (3) plan options and a 14-Day Trial: Basic Free, Standard, and Pro version

Simulcast to YouTube and Facebook

KEY FEATURES INCLUDE:

- Use Be Live in Chrome Browser
- Schedule Live Webcast
- Add Guests or Co-Host
- Guest Labels: Name
- Add Logo, Background, Text
- Allows viewer Comments
- Screen share
- Add Agenda
- Changing view mode
- Social Media comments

CHAPTER 7

LAUNCH AN ONLINE CHURCH (ECHURCH)

Then He said to them, "Follow Me, and I will make you fishers of men."
MATTHEW 4:19

Welcome to the Information Age. With the world wide web/internet, no one should ever miss one of your Worship Services, a Bible Study, life changing messages, or ministry events.

Broadcast your church sermons, men, women, and youth ministry events worldwide:

- Weekly Church Services
- Bible Study
- Men, Women, & Youth Ministries
- Church conferences and events

DID YOU KNOW?

USA Churches classifies churches in our directory by size – mega churches, large churches, medium churches, and small churches – and churches come in all sizes.

SOURCE: USACHURCHES.ORG

MEGA CHURCH
Average weekend attendance more than 2,000 people

LARGE CHURCH
Average weekend attendance between 301 and 2,000 people

MEDIUM CHURCH
Average weekend attendance between 51 and 300 people

SMALL CHURCH
Average weekend attendance 50 or fewer people

WHAT IS AN INTERNET MINISTRY OR INTERNET CHURCH?

LIVE GLOBAL WORSHIP EXPERIENCE

With the integration of Internet, Social Media, and Live Streaming, houses of worship have unlimited possibilities to increase their outreach, financial support, and sharing the glorious Gospel of Jesus. There are plenty of churches who have not taken advantage of live streaming their services. Some of the concerns of churches is the possible negative impact on church attendance, costs of equipment, staff, or volunteers to run the media department.

With the proper training, preparation, and team, an online church can be very beneficial to churches. It is definitely something to prayerfully consider because it is a time-consuming commitment and everyone needs to be on the same page.

> ## "BIZ TIDBIT
>
> Churches can harness the power of the internet and create an Online Church as a virtual extension of their ministry outside of the brick and mortar [buildings].
>
> **SHERRY BRONSON**"

DIGITAL TERMINOLOGY

- eChurch is the digital name for the Internet Church Campus. Individuals worldwide receive a virtual worship experience directly from your dynamic website.

- eMembers is the digital name for individuals who view weekly church services online. eMembers actively participate in church activities via the internet. They pay their tithes, make donations, salvation prayer, prayer requests, and purchase ministry products, etc.

- eTithing: eMembers are able to pay their tithes electronically online 24 hours a day on your church website.

DEFINITION

What is an internet ministry? Internet ministry refers to a wide variety of ways that a religious group is using the internet to facilitate its religious activities, particularly worship services. Referred to also as an Online Church.

GETTING STARTED:
LIVE STREAMING WORSHIP SERVICES

PRODUCTION PLANNING AND SCHEDULING

Pre-production planning meetings are definitely required for the media team staff and/or volunteers. Communication is crucial and having an actual plan of action. The online church media team will need to have a copy of the live production schedule in advance.

Things to consider for professional live streams

- Do you want graphics such as titles/lower thirds, Bible verses, overlays?
- **B-ROLL:** Will you have church announcements prior to or during your Live webcast?
- Will the church promote ministry products and upcoming events during the live stream?
- **MULTI-CAMERA PRODUCTION:** Do you want to capture audience video shots?

Why?	Who?	What?	When	How?	Where?
• Purpose	• Define Target Audience	• Product	• Daily	• Gear	• Onsite
• Profit		• Services	• Weekly	• Technology	• Remote Locations
• Promotion			• Bi-Weekly	• Volunteers	• Home
• Awareness				• Hire Staff	
• Charity				• Social Media Manager	

Pre-production planning meetings should include the following:

1. Online Church official launch date
2. Live Streaming Broadcast Schedule
3. Talent Releases (guest recording artists or guest speakers)
4. Copyright & License for Music (use royalty music)
5. Video Production Plan (Pre, Production/live, & Post Live strategy)

ONLINE CHURCH WEBSITE

LIGHTS, CAMERA, & GO GLOBAL!

The great advantage of the internet is the ability to host events virtually from anywhere. There is a cost-effective way to launch an Online Church using a Website. Creating professional looking websites is easier than ever. Please ensure that the website is responsive to be viewable on mobile devices such as an iPhone, Android, and iPad. There are many website design sites that allow us to create websites without being a technical expert. Wix.com has an easy learning curve if the ministry cannot afford to hire a professional website designer. Wix.com offers affordable website plans and their website templates can be edited to fit the church brand.

Another option to build a website is to use WordPress.com. WordPress also offers website templates that can be edited and there are free plugins to help with a more professional website. Be sure to research all options before making a decision to build your own website.

Visit www.Wix.com and WordPress.com to learn more about hosting and a domain.

Online Church: Small and Rural Churches – Facebook Live Option

Churches with limited budgets do not need a website to launch an online church. For small and rural churches, please use a Facebook Business Page to launch your online church. Churches can go live directly to their Facebook Business. When setting up your Facebook Page, please sign-up as a non-profit and add a donate button to your page for members to pay tithes, offerings, and collect general donations.

STARTER WEBSITE PAGES

Responsive Website for Mobile Device Friendly (iPhone, android, iPad, tablet).

1	About Us, Mission and Vision
2	**HOME PAGE:** Live Streaming: Embed Video Media Player – Facebook
3	**HOME PAGE:** Add social media icons and hyperlink Facebook, Twitter, YouTube etc. Web designer can also add a Facebook plug-in on to your page
4	**DONATE PAGE:** Tithes, offerings, and general donations
5	**NEW & EXISTING MEMBERS:** Welcome, Baptism, etc.
6	**ONLINE STORE (ESTORE):** Sell MP4, MP3, DVDs, eBooks, Books, etc.
7	Contact Us

Recommended - Custom Website Designers

- CTSGraphics.net
- KingdomChurchWebsites.com
- SquareSpace.com
- WordPress.com
- Wix.com

GETTING STARTED:
LAUNCH AN ONLINE CHURCH

LIVE STREAM GEAR SETUP
Windows only

It's time to share the Gospel of Jesus Christ worldwide. Execute your pre-production plan and please celebrate the small victories. Churches should already have some of this gear already, such as a computer, microphones, audio-mixer or console, church presenter software to display scriptures, music lyrics, sermon message titles, etc. Churches could get started with $1,500.00 or less.

Technical: When purchasing the live video streaming software, please review the vendor system requirements for the computer to run their software: Operating systems, processor, memory (RAM) hard drive, and graphics card needed, etc.

Gear / Technology: Windows Computer (laptop or desktop)

- **OPTION 1:** Camera Camcorder or PTZ Optics
- **OPTION 2:** DSLR or Mirrorless Camera (HDMI Clean output/ adjust settings to remove display)

Live Streaming Gear Set-up
Windows PC

Camera Camcorders (options)

Capture Cards (3 options)

LIVE STREAM

ProPresenter.

LIVE VIDEO PRODUCTION SOFTWARE

Wirecast   vMix

Audio Mixer

LIVE STREAM GEAR SETUP
Mac PC

TECHNICAL: When purchasing the live video streaming soft-ware, please review the vendor system requirements for the computer to run their software: Operating systems, proces-sor, memory (RAM) hard drive, and graphics card needed, etc.

Gear/Technology: Mac PC

- OPTION 1: Camera Camcorder or PTZ Optics
- OPTION 2: DSLR or Mirrorless Camera
- CAPTURE CARD: Blackmagic Design Ultrastudio Mini Recorder, Elgato Cam link 4K
- LIVE STREAMING SOFTWARE: Wirecast, vMix or OBS
- OPTION 1: BEHRINGER UM2 Audio Interface
- OPTION 2: BEHRINGER Audio Mixer
- MICROPHONE(S): Handheld, Wireless
- Lighting
- OPTIONAL: Church Presentation Software (license)
- ProPresenter.com (NDI out) or EasyWorship.com
- OPTIONAL: Video Switcher (Multiple Cameras)
- Blackmagic Design ATEM Mini Live Production Switcher $295
- Roland VR-1HD AV Streaming Mixer $1,095

Live Streaming Gear Set-up
MAC PC

LIVE STREAM

Camera Camcorder

Capture Card

ProPresente

ecamm
Live Stream software for MAC only

Audio Mixer

CHAPTER 8

RESOURCES

Here is a list of various vendors to purchase gear, technology, and services.

LIVE STREAM SERVICES & PLATFORMS SHARE YOUR MESSAGES

- ChurchStreaming.tv (live stream services)
- Sermoncast.com (live stream services)
- LiveStream.tv (share your messages)

VIDEO CAMERA CAMCORDER, DSLRs, MIRRORLESS CAMERA

- Sony.com
- Canon.com
- Panasonic.com
- Amazon.com
- BHphotovideo.com

SPARKOCAM VIRTUAL WEBCAM SOFTWARE

- Website: https://sparkosoft.com/sparkocam
- Use Canon or Nikon DSLR Cameras as a regular webcam
- Supported platform: Windows only. Mac isn't supported.

VIDEO EQUIPMENT, AUDIO, COMPUTERS, CAPTURE CARDS & ACCESSORIES

- CowboyStudio.com
- CheapLights.com
- Bhphotovideo.com
- Kingdom.com
- Rode.com

- GuitarCenter.com
- Elgato.com (capture cards)
- AJA.com (capture card)

VIDEO EDITING SOFTWARE & TOOLS

- Filmora.wondershare.com
- Adobe.com – Adobe Premiere Pro CS6
- SonyCreativesoftware.com – Sony Vegas Pro
- Apple.com (Final Cut Pro X, video editing software)
- DigitalJuice.com (video production tools)

LED AND FLUORESCENT, SOFTBOX, RING LIGHTING

- Fovitec.com
- Amazon.com
- BHphotovideo.com
- iOgrapher.com
- Theringlightstore.com

LIVE STREAM VIDEO CONFERENCES ON FACEBOOK

- Zoom.us
- Webex.com
- Uberconference.com
- ClickMeeting.com
- GotoMeeting.com

CREATE & SELL AN ONLINE COURSES

- Teachable.com
- Udemy.com
- Skillshare.com
- Kajabi.com

STREAMING MEDIA TERMINOLOGY

- https://www.streamingmedia.com/Glossary/

SMS MOBILE MARKETING
EZTexting.com

- Free Account or Paid Account
- Register to their Mobile Phone Number (Embed into your website)
- Collect & Communicate with New & existing contacts
- Text Message updates and reminders
- Donations

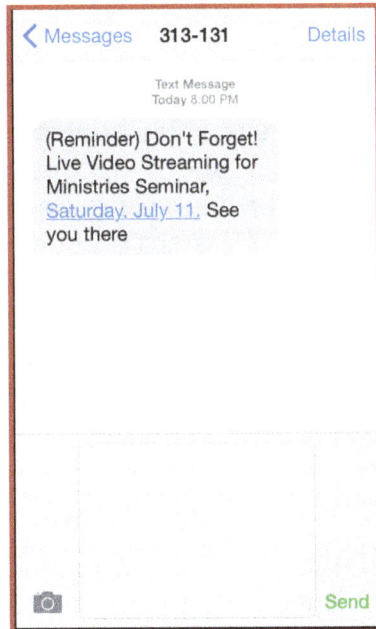

ABOUT THE AUTHOR

SHERRY BRONSON

Sherry Bronson is an outstanding television host, television producer, live video streaming consultant, motivational speaker, and author. She delivers powerful life changing messages that speak to the heart of any audience. Her relentless passion to see individuals' lives changed shines through her energetic, compassionate, and magnetic presence.

As founder of Bronson Media, she teaches media workshops and webinars to equip and empower small businesses and houses of worship, and ministries with the tools needed to expand their reach through video and live streaming.

With over 15 years of media experience, including television studio production, live video streaming services, talk show host, television producer, she is making great strides in media. As a leader, she encourages her team to embrace change, a collaborative spirit, positive attitude, and continue to expand their skills and abilities.

HUMBLE BEGINNINGS: SHERRY BRONSON'S STORY
OVERCOMING ADVERSITY

Ms. Bronson is a true testament that the power of faith, courage, perseverance, and hard work really do pay off. From her humble beginnings, the odds of escaping an impoverished life and becoming a success story were truly near impossible. She was raised in the home of a single mother of twelve children, dysfunctional home life, and absentee father, struggled with rejection, poor self-image, and low self-esteem. In her early teen years, she went from the comfort of a 4-bed room bricked home to sleeping on the floor of an alcoholic. Between the ages of 11-14, the family moved over 6 times. The unstable environment greatly increased the chances of her becoming a high school dropout or another at-risk teen statistic. As a young teen girl, she reflected on her life and witnessed much family heartache, generational poverty, and abuse. She was confronted with the hard-core reality that her chances of personal and professional success were slowly fading. With great determination and courage, Ms. Bronson put a stamp of rejection on poverty and generational dysfunction.

EDUCATION: A CRITICAL KEY TO BREAKING THE CYCLE OF POVERTY

She knew completing her educational studies was one of the keys to success. In the senior years of high school, she found stability in the home of an older sister who counseled, encouraged, and supported her through those years until she graduated. Ms. Bronson went on to college and was the first in her family to receive a bachelor's degree. She graduated from Jackson State University, Jackson, Mississippi in 1993. She embarked on a new journey of self-discovery and fulfilling her divine purpose and destiny.

PURPOSEFUL LIFE

She has been chosen as a beacon of hope for women and men. Ms. Bronson has a natural gift for inspiring and motivating individuals to never stop believing in themselves. With her high energy, magnetic personality, and passion for changing lives, she continues to draw a diverse following.

She relocated from Jackson, Mississippi to Dallas, Texas to discover and fulfill her divine purpose. It was a giant leap of blind faith. She had less than $100, no job, and a dream in her heart to live a life of success. On her new journey, she carried in her heart the most important inspiration, and that was "A Word of Promise from the Lord." God did not fail her. As it is said, the rest is history.

Through television, speaking, and writing, she is empowering individuals to draw from their God-given power within to realize their dreams. She studies the Bible and understands that faith comes by hearing and hearing by the word of God. She encourages individuals not to focus on past shame, guilt, failures, and disappointments. She believes that God created us with divine purpose and destiny.

Ms. Bronson encourages individuals to draw from their God-given power within to make their dreams a tangible manifestation.

Sherry Bronson oversees all facets of the business launch and success from the ground up. Provides live video streaming consulting services and organizes media seminars for small businesses, professional organizations, communicators, houses of worship, and ministers.

Interact with client through every phase of the television show production to incorporate feedback and implement suitable suggestions while focusing on exceeding clients' expectations. Discuss changes within TV show production set design prior to execution. Deliver services of a director along with hiring production crew, editing episodes, and broadcasting on ROKU streaming channel. Create and implement online social media strategies and plans in close collaboration with clients.

Managed streaming of television production projects from development to postproduction: clients were The DeNesha Rachelle Show (Roku), Queendom Talk with Marlo Mozee (Roku), and Triumph Ministries with Ramona Long (Roku) within allotted budget and time.

Conduct various seminars and webinars on variety of new topics, such as How to Launch Your Television Talk Show, How to launch an online church on a Shoestring Budget, 4 C's the Business of Live Streaming: Content, Capture, Convert, and Cash, On-Camera Workshop: Lights, Camera, and Speak, and How to Launch Your Own Roku Television Channel.

Professional Associations/Memberships/ Recognition

- 2014–2016 Christian Media Association, Dallas Chapter Leadership Team

- Conference Speaker: Christian Media Association National Conference

- 2005–2009 member, Dallas Community Television/Dallas iMedia

- 2010–present, member, The Professional Woman Network

- 2017–present, member, Christian Women in Media Association (CWIMA)

- 2018 Community Service Award, National Association of Business & Professional Women's Clubs

JOIN SHERRY BRONSON ON FACEBOOK LIVE & QA

Insider Pro Live Video Streaming Tips & Tutorials
Facebook.com/BronsonDigitalMedia

Email: info@BronsonDigitalMedia.com
Web: www.BronsonDigitalMedia.com
Web: www.SherryBronson.com

PRAISE FOR BRONSON MEDIA
WORKSHOP: LIVE VIDEO STREAMING

"This interactive class is a must for anyone who is interested in live streaming and media, I would encourage you to come out and attend one of Bronson Media's magnificent classes."

~ TIME WITH THERESA WESTBROOK, TELEVISION TALK SHOW & TV PRODUCER

"It's been a great inspiration, I attended the Live Video Streaming for Ministries 101 Seminar, Sherry Bronson is phenomenal. I learned so much as well as my soul was blessed. If anyone is interested in moving their ministry to the next level in this technology age; this seminar is a must."

~ ROCHELLE OWENS, CHRISTIAN TEMPLE WORSHIP CENTER

"Excellence, Integrity, and Professionalism in Media"

~ TIM SHIELDS, CO-DIRECTOR OF CHRISTIAN MEDIA ASSOCIATION (CMA)

"What I have learned from Sherry Bronson with live streaming, it is so important to know that there are other options out here when it comes to growth of your ministries so I really learned a lot today. I advise anyone that needs to know more about live streaming contact Sherry Bronson to help grow your ministry."

~ PASTOR DENISE WOLFORD, BREAD FROM HEAVEN CHRISTIAN CENTER

"Holy Arm Ministries are pleased with the service that Bronson Media provides for all of our live streaming events. The service is professional and of great quality. "

-PAULA L. SMITH, CEO AND EXECUTIVE PASTOR, HOLY ARM MINISTRIES, INC.

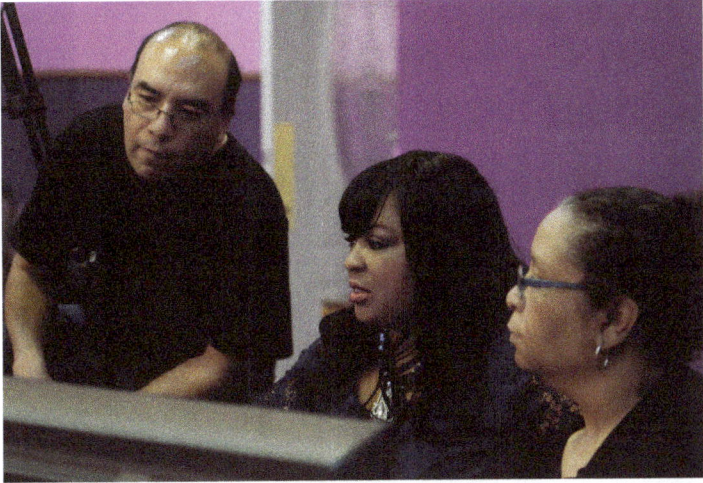

www.ingramcontent.com/pod-product-compliance
Lightning Source LLC
Chambersburg PA
CBHW071440210326
41597CB00020B/3878